Encounter Earth

Interactive Geoscience Explorations

STEVE KLUGE

Upper Saddle River, NJ 07457

Editor-in-Chief, Science: Nicole Folchetti
Publisher, Geosciences: Dan Kaveney
Acquisitions Editor: Drusilla Peters
Media Editor: Andrew Sobel
Marketing Manager: Amy Porubsky
Managing Editor, Science: Gina M. Cheselka
Media Production Manager: Rich Barnes
Cover Designer: Suzanne Behnke
Compositor: Thomas Benfatti
Senior Operations Supervisor: Alan Fischer

Pearson Prentice Hall
Pearson Education, Inc.
Upper Saddle River, NJ 07458

Printed in the United States of America

10 9 8 7 6 5 4 3 2 1

ISBN-13: 978-0-321-58129-7

ISBN-10: 0-321-58129-6

Pearson Education Ltd., *London*
Pearson Education Australia Pty., Ltd., *Sydney*
Pearson Education Singapore, Pte. Ltd
Pearson Education North Asia Ltd., *Hong Kong*
Pearson Education Canada, Ltd., *Toronto*
Pearson Educación de Mexico, S.A. de C.V.
Pearson Education—Japan, *Tokyo*
Pearson Education Malaysia, Pte. Ltd.

Contents

Exploration 1:
Using Google Earth™ Mapping Service

In order to use these explorations, you must have Google Earth™ mapping service software installed on your computer. The program is available in several versions, including the free one for which these explorations are designed, at http://earth.google.com. Running the program requires a high-speed internet connection, and you will find navigating is easier with a mouse than it is with a touch pad and keyboard, though either will work.

Getting Started

Go to http://earth.google.com and download and install the latest version of the Google Earth™ mapping service software (Free) on your computer.

Click the "Product Tour" link, and view the various tutorial videos. Pay particular attention to the following, as they are important to getting the most out of these explorations:
• Your World in 3D – Zoom, Tilt, and Rotate
• Layers of Mapping Information

Then, PLAY! Open the Google Earth™ mapping service software and fly around the globe, find your campus, zoom in to your dorm. Fly to your home and to your favorite vacation spots. Give yourself some time for this—it will familiarize you with the controls, and reduce the distraction of play when you do get to work on these Explorations.

Encounter Earth: Interactive Geoscience Explorations

Log on to the *Encounter Earth* site – http://www.mygeoscience.com/kluge – and click the link for the "Exploration 1: Using Google Earth™ Mapping Service" KMZ file to open that file in the Google Earth™ mapping service program. (This is the same file you'll see later in Exploration 11: Plate Tectonics—Divergent and Transform Boundaries.) You can use it to follow the figures and instructions that follow.

Figure 1.1 shows the "sidebar" that appears when you first open the Google Earth™ mapping service program. There are 3 panels in the sidebar: Search, Places, and Layers.

Each of these panels can be expanded or hidden by clicking the triangular icon next to the panel label.

You will not typically need the Search feature while working on these Explorations, so you can click the triangular icon to hide it right away.

The Places panel is where most of the action takes place. When you click a KMZ file to open it, it will appear in the Temporary Places folder within the Places panel. In Figure 1.1, the Plate Tectonics—Divergent and Transform Boundaries exploration has been loaded.

You can work from the Temporary Places folder, or you can right-click content in Temporary Places to save it to the My Places folder (it really doesn't matter; you will be given the option of saving your Temporary Places content to My Places when you exit at the end of your session). Content in My Places will be saved when you close the application, and will appear there the next time you open it.

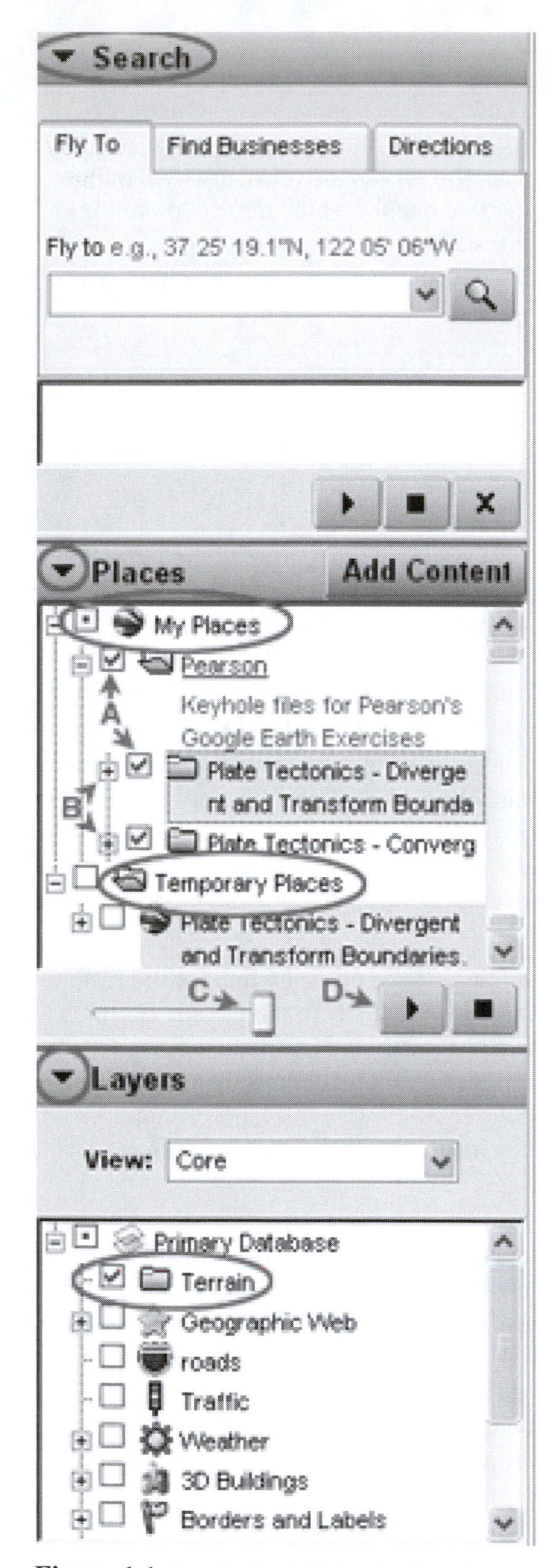

Figure 1.1 Google Earth™ is a trademark of Google, Inc.

Once you have downloaded a KMZ file, there are several controls used to manage the display of the file contents.

Refer to Figure 1.1, and notice the following controls:

A - Checking or unchecking the boxes will display or hide the content.
B - Content folders can be expanded or hidden by toggling between + and – in the boxes next to the folder name.
C - Some content, like overlays and paths, can be made more or less opaque with this slider.
D - The view will automatically fly from placemark to placemark within a folder when the play button is clicked.

The Layers panel contains all sorts of "built-in" content, and you should take the time to browse those various layers. They can be turned on and off by clicking the rectangular boxes next to the layer names. You will definitely want the Terrain layer turned on as you work through these Explorations (it provides the 3D view of the Earth), but you should turn off all the other layers to avoid cluttering the views as you work.

Once you have turned Terrain on and the other layers off, you can click the triangular icon next to Layers to hide that panel, too.

Content in *Encounter Earth*

Notice that the Search and Layers panels have been hidden in Figure 1.2, leaving the entire sidebar for the Places panel. You will probably find it easier to work through these Explorations with your sidebar set this way, too.

There are several types of content in these Explorations:

Each Exploration will download as a .kmz file (A in Figure 1.2). Within that file there will be one or more folders (B and C in Figure 1.2), and each folder will contain some combination of placemarks, paths, shapes, and overlays (D in Figure 1.2).

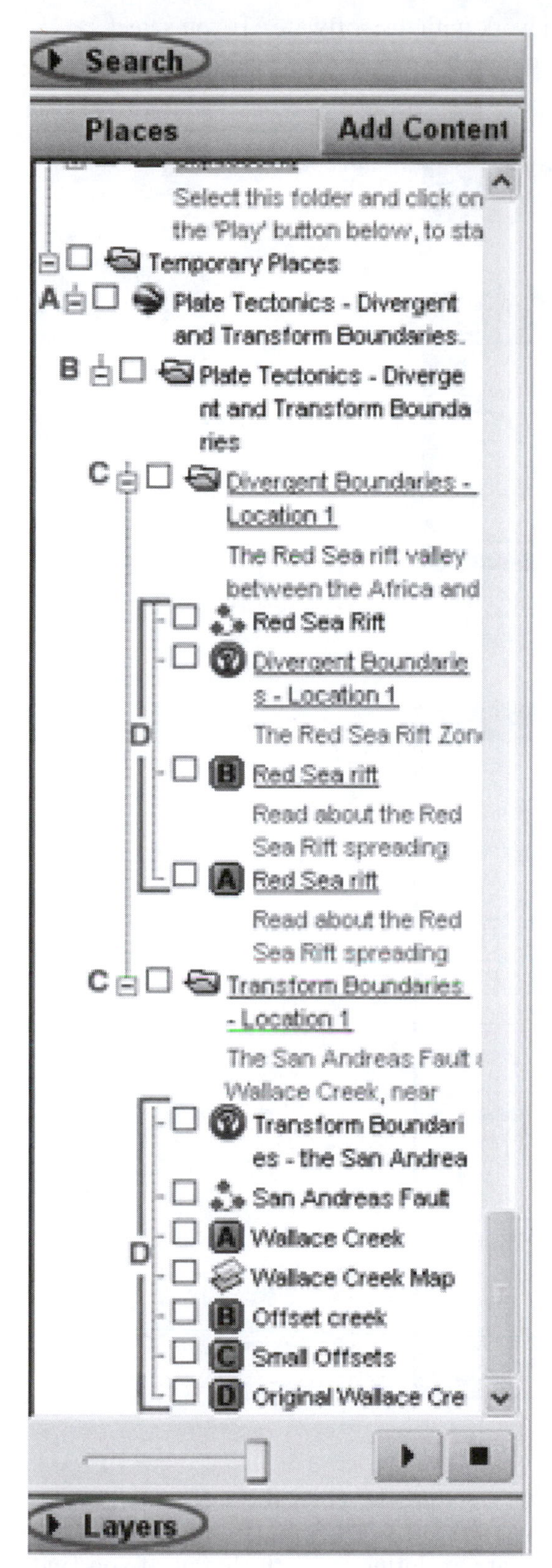

Figure 1.2 Google Earth™ is a trademark of Google, Inc.

Placemarks are content that is locked to a particular place on Earth's surface. They may simply mark a place, or they may contain some combination of text, hyperlinks, and images. A placemark can be viewed by clicking it in the sidebar, or by clicking its icon in the main display.

Each investigation area will be marked with a placemark having a "?" icon, as indicated in Figure 1.2. Specific placemarks within the investigation area will be assigned other icons.

A path is simply a line or route drawn on the main display. In Figure 1.2, the Red Sea Rift and the San Andreas Fault are traced with paths.

A shape is a flat geometric shape that can be displayed at any elevation in the main display.

An overlay is a map or other image that is draped over the surface of the main display. An important feature of overlays is that the opacity of the overlay can be adjusted. This feature is very helpful when comparing the appearance of a landscape with a map image of it.

Other tools and controls

Google Earth™ mapping service (Free) contains a number of useful tools that you will use as you Explore. Click "View" in the toolbar at the top of the Google Earth™ window, and click the listed features on and off each to see what they do. Make sure that at least the Status Bar is turned on. It provides latitude and longitude and elevation information at the bottom of the main display.

You will use the ruler frequently as you work through these investigations. It is accessed by clicking "Tools" in the top toolbar and selecting "Ruler" from the drop-down menu (Figure 1.3).

Notice that you can select from a variety of measurement units, and that you can measure the length of a path as well as make simple point-to-point measurements.

You will also find it necessary to make other adjustments as you work with the software. If you select "Options" from the Tools dropdown, you are presented with this dialog box (Figure 1.4). (Mac users will find this feature under Google Earth > Preferences.)

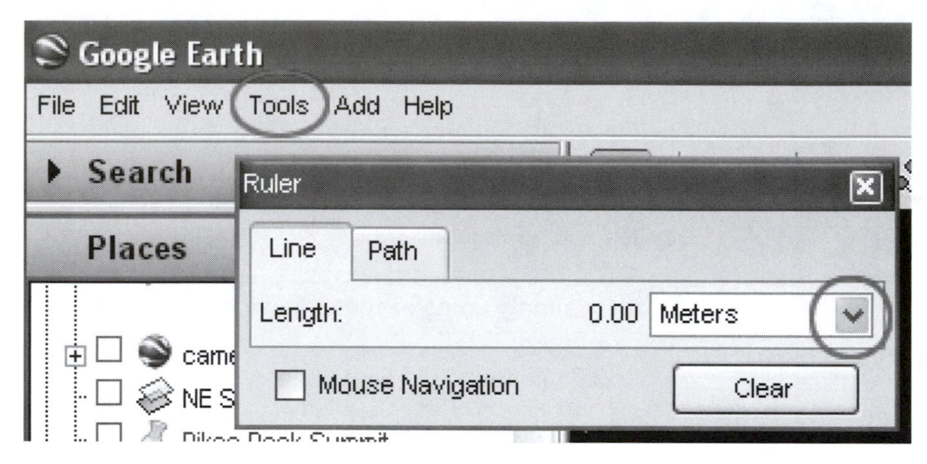

Figure 1.3 Google Earth™ is a trademark of Google, Inc.

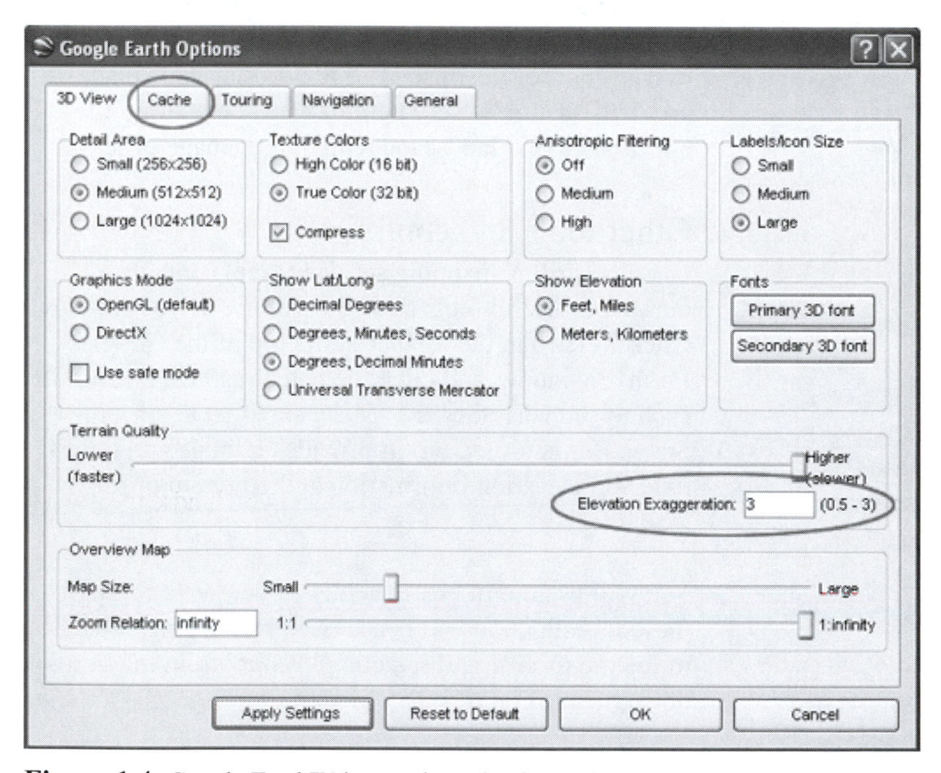

Figure 1.4 Google Earth™ is a trademark of Google, Inc.

Generally, the default settings will work fine, but you may find that resetting the vertical exaggeration is necessary to get the most out of the displayed landscape. Use a setting of 0.5–1 when zooming in close on areas of high relief, like mountains, and a setting of 3 when the area you are exploring is relatively flat. If you find that the application begins to slow down, you may want adjust your cache size, and to get a clean start, clear your caches. You access those controls by clicking the "Cache" tab in the Options dialog box.

Exploration 2:
Interpreting Topographic Maps

Grand Canyon, Arizona

Log on to the *Encounter Earth* site – http://www.mygeoscience.com/kluge – and click the link for the "Exploration 2: Interpreting Topographic Maps" KMZ file to begin this activity.

In the Google Earth™ mapping service, expand the Interpreting Topographic Maps folder and click on the Grand Canyon Map overlay. Set the opacity slider about halfway along its length. This will allow you to see the actual land surface beneath the map. As you work through this Exploration, you may want to return to this slider to change the opacity to suit your needs.

1. Double click the icon for placemark 1 in the sidebar to fly to that view. Answer the following questions.

a. Notice that every fifth contour line is drawn more boldly than the others, and that some of them are labeled (here, the 5400 and 6000 ft contours are labeled (placemarks 1 and 2 respectively, and there are 2 unlabeled bold contours between them). Bold contours are called ***index contours***. On this map, how many feet are there between index contours?

________________ft

b. Notice that there are 4 "regular" contours drawn between successive index contours, leaving 5 spaces, or ***contour intervals***, between index contours.

c. Calculate and report the change in elevation across each contour interval (that is, between successive "regular" contours) on this map.

________________ft

Depending on the scale of the map and the relief of the landscape, mapmakers will use contour intervals as low as 5 ft on very flat landscapes, and as high or higher than 100 ft on steep, rugged landscapes.

2. Double click the icon for placemark A in the sidebar to fly to that view.

a. Describe the topography at A, and notice and describe the spacing of the contour lines at A.

b. Describe the topography at B, and notice and describe the spacing of the contour lines at B.

c. Make a statement about the spacing of contour lines relative to the slope of the land.

3. Double click the icon for placemark C in the sidebar.

a. Study the slope of the land at placemarks C, D, and E. Describe how the spacing of the contour lines relates to the slope of the land at each placemark.

4. Double click the icon for placemark F in the sidebar.

a. Turn off the map overlay by unchecking its box in the sidebar. Use the path tool (from the toolbar at the top of the display) to draw a path between the two placemarks F in such a way that the elevation of the path is constant. Make your points along the path short, doing the trip from F to F in 10 or 15 steps.

Once you've drawn your path, turn the map back on, and compare what you've drawn with the contour lines on the map. Does your path nearly parallel the contours that cross the valley?

b. Notice the stream that flows NW from the top of the screen and between placemarks F. Describe what happens to the contour lines where the stream cuts across them. You may want to use a letter of the alphabet to describe the shape of the contour lines.

c. Which of the following correctly describe(s) the direction in which the "V"s formed by the contour lines point as a stream crosses them? (circle each correct response)

uphill	downhill
north	south
upstream (opposite the flow of the stream)	downstream

d. Browse the map to find other streams, and verify that the answers you reported in c. above apply to other streams as well.

e. Locate a valley without a stream flowing in it. Do the contours generally behave the same way as they do where streams cross them?

5. Double click the icon for placemark G, and notice that placemarks G and H lie along the Bright Angel Trail (the dotted line on the map).

a. The gradient of a surface is a way of expressing the steepness of the landscape by reporting the change in elevation that occurs over some horizontal distance. The gradient can then be easily calculated by dividing the rise in the landscape by the horizontal run required to gain that rise.

Using the Status Bar at the bottom of the display, determine the elevation of placemark G ____________

and placemark H ___________ and calculate the rise, or change in elevation between them.

Change in elevation, or rise, from G to H = ___________ft

Now use the Ruler tool to measure the straight line distance, or run, from G to H in miles.

Straight line distance, or run, from G to H = __________mi

Reset the ruler and click the Path tab to measure the distance from G to H along the trail in miles.

Distance along the trail, or run, from G to H = ___________mi

Now calculate the straight line gradient from G to H (rise/run), noticing that the units of your answer will be ft/mi.

Straight line gradient from G to H = ____________ft/mi

Calculate the gradient from G to H along the trail.

The gradient along the trail from G to H = _________ft/mi

Describe the benefits and tradeoffs a hiker makes in following the trail, rather than scrambling straight up the slope, from G to H.

6. Double click the icon for placemark I on the sidebar.

a. Describe the topographic feature at placemark I, and describe the appearance of the contour lines on this feature.

b. How is the appearance of the contour lines on this ridge similar to the appearance of contour lines in a valley?

c. What is the important difference in the appearance of the contour lines on a ridge?

7. Double click the icon for placemark J on the sidebar.

a. Describe the landform at placemark J.

b. Describe the appearance of the contours at placemark J.

c. Notice the X drawn on the map at the very top of The Battleship. How closely does the labeled elevation of 5850 ft match the elevation in the Status Bar at the bottom of the display?

8. Double click the "?" icon in the sidebar to re-center your view, and then zoom into the SE corner of the map at an eye elevation of about 13,500 ft.

a. Find the "BM x 6973" just east of the Visitor Center (at latitude 36° 3.380' N, longitude 112° 7.239' W). The "x" marks a benchmark, a reference point of known elevation and location marked with a bronze disk permanently mounted in the ground there. A typical benchmark is shown in Figure 2.1.

Scan the map for another benchmark, and report the elevation and location of it here:

Elevation = ___________

Location = latitude _______________

longitude _______________

Figure 2.1

9. Double click the "?" icon in the sidebar to re-center your view. Use the opacity slider to quickly change the map back and forth from opaque to transparent.

a. Which was updated more recently, the map or the satellite imagery, and how do you know that?

b. Describe the changes in the land use at Grand Canyon for the period of time between when the map was made and when the satellite imagery was made.

Exploration 3:
Running Water — Stream Dynamics

Lawn Lake, Roaring River, and Fall River, Rocky Mountain National Park, Colorado

Log on to the *Encounter Earth* site – http://www.mygeoscience.com/kluge – and click the link for the "Exploration 3: Running Water—Stream Dynamics" KMZ file to begin this activity.

The Lawn Lake Flood

Open the "Lawn Lake Flood" folder and double click the "?" icon next to the Lawn Lake Flood placemark to fly to it. Open the placemark balloon and read about the cause of the flood.

1. Double click the icon for placemark A.

a. Using data from the Status Bar at the bottom of the display, calculate the change in elevation between placemarks A and B. Record your answer in meters.

Elevation of placemark A = _______________m

Elevation of placemark B = _______________m

Change in elevation between placemarks A and B = __________________m

b. Use the Ruler tool to measure the length of the channel of the river between placemarks A and B. Do not measure the straight line path—use the path tool instead. Record your answer in km.

Length of the river channel = _______________km

c. Calculate the gradient of the stream channel between placemarks A and B by dividing the change in elevation by the length of the channel.

Gradient of stream channel between placemarks A and B = ________________________

2. Double click the icon for placemark B to zoom back in for a closer view.

a. Repeat the steps used above to calculate the gradient of the valley floor between placemarks B and C.

Elevation of placemark B = _______________m

Elevation of placemark C = _______________m

Change in elevation between placemarks B and C = __________________m

Gradient of stream channel between placemarks A and B = ________________________

b. Double click the icon for placemark D in the Places panel. Use the Ruler tool to determine the size of the largest boulders in this field of view.

c. Double click the icon for placemark C in the Places panel, and compare the size of the sediments at placemark C with those at placemark B.

d. Describe how the sudden change in gradient at location B was responsible for the rapid deposition of the alluvial fan here, and why the sediments are smaller and finer at placemark C.

3. Double click the icon for placemark E in the Places panel.

a. Cite whatever evidence you can to support the argument that sediments at placemark C were deposited before those at placemark E.

b. Describe the effects of the current deposition at placemark E on Fall River, which flows into this area from the west. (You can double click the Fall River icon in the Places panel for a wider view.)

4. Double click the icon for the Meander Map Area in the Places panel. Set the elevation display in the Status Bar to "Feet, Miles" by clicking Tools > Options > 3D View tab.

a. Figure 3.1 is a map of a section of the Fall River corresponding to the Meander Map Area shape in the Places panel. Draw in the point bars and label the cut banks on the meanders on Figure 3.1. Describe where in the meanders each occurs, and explain how and why they form where they do.

b. Follow the procedure followed in step 2 above to determine the following gradients (note that the unit of your gradients will be feet/mile):

Gradient along the stream channel from placemark F to placemark G = ________________

Work:

Straight line gradient from placemark F to placemark G = ________________

Work:

c. Imagine that during a period of peak flow, the river briefly breaches its banks, and a small channel is cut directly across the land between placemarks F and G. Consider the answers you provided in 4b, and describe and explain the processes by which the meander between placemarks F and G could be cut off from the main flow of the river.

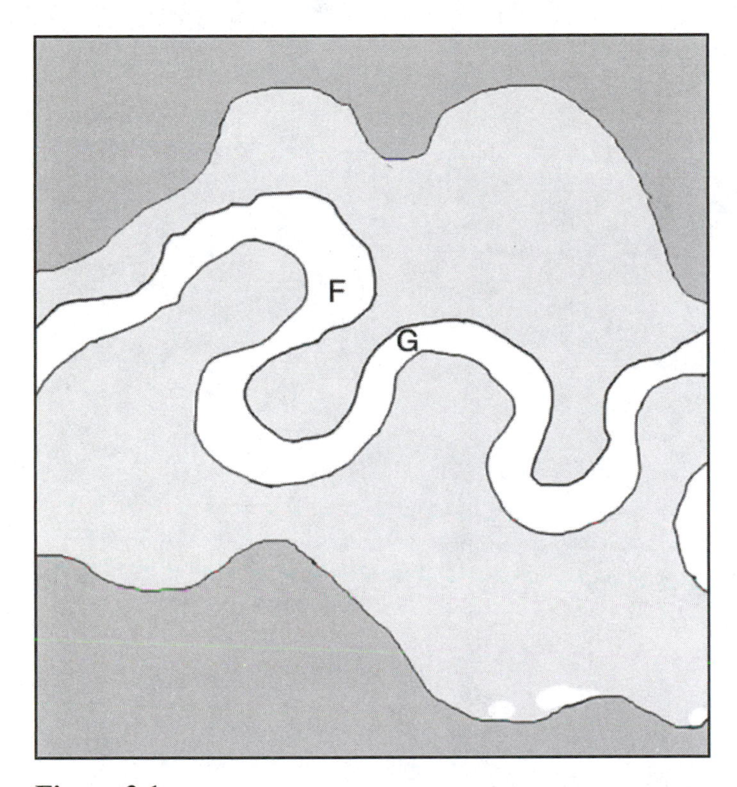

Figure 3.1

Exploration 4:
Deposition — Angle of Repose

Huerfano River valley, Sangre de Cristo Mts.
Colorado Flatirons, near Boulder, Colorado
Crater Lake, Oregon
Mt. Bachelor, Oregon
Mauna Kea, Hawaii

Log on to the *Encounter Earth* site – http://www.mygeoscience.com/kluge – and click the link for the "Exploration 4: Deposition—Angle of Repose" KMZ file to begin this activity.

Angle of Repose—Location 1: Huerfano Valley, Colorado.

Click the icon for this placemark to fly to it, and view the image in the placemark balloon. You are looking south along the Huerfano River valley in the Sangre de Cristo Mountains of southern Colorado. Many of the slopes on the sides of the valley are formed of loose, unconsolidated material that has tumbled from the peaks above. The angle of repose labeled on the image was measured with a protractor placed directly on the image.

1. Turn off the placemark balloon and familiarize yourself with the landscape by flying around a little. Then use the Ruler tool and elevation data in the Status Bar across the bottom of the display to calculate the angle of repose of the material making up the slope near the placemark (angle of repose = arctan (elevation /distance)).
Work:

Calculated Angle of Repose = _____________

How does your calculated angle of repose compare with the measured angle in the placemark image?

Angle of Repose—Location 2: Flatirons, near Boulder, Colorado.

Click the icon for this placemark to fly to it, and view the image in the placemark balloon. The view is up a talus slope along the trail to Royal Arch in the Flatirons west of Boulder, Colorado.

1. Close the placemark balloon and familiarize yourself with the landscape by flying around a little. Then use the Ruler tool and elevation data in the Status Bar across the bottom of the display to calculate the angle of repose of the material making up the slope around the placemark (angle of repose = arctan (elevation / distance)).
Work:

Calculated Angle of Repose = _____________

How does the angle of repose here compare with the angle of repose on the talus slopes in the Huerfano valley?

Angle of Repose—Location 3: Crater Lake, Oregon.

Placemark A. Open the Crater Lake, Oregon folder and click the icon for placemark A to fly to it and view the image in the placemark balloon. The view is southeast into the eroded core of Mt. Scott, an old volcano just east of Crater Lake in Oregon.

1. Close the placemark balloon, and use the Ruler tool and elevation data in the Status Bar across the bottom of the display to calculate the angle of repose of the material making up the slope at placemark A (angle of repose = arctan (elevation /distance)).
Work:

Calculated Angle of Repose = ____________

How does your calculated angle of repose compare with the angles you calculated for the Huerfano valley and the Flatirons?

Placemark B. Click the icon for placemark B to fly to it and view the image in the placemark balloon. The view is southeast into the eroded core of Mt. Scott, an old volcano just east of Crater Lake in Oregon.

1. Close the placemark balloon, and use the Ruler tool and elevation data in the Status Bar across the bottom of the display to calculate the angle of repose of the material making up the slope at placemark B (angle of repose = arctan (elevation /distance)).
Work:

Calculated Angle of Repose = ____________

How does your calculated angle of repose compare with the angle you calculated for the bare slope at placemark A?

Placemark C. Click the icon for placemark C to fly to it and view the image in the placemark balloon. The view is northwest along the southern flank of Mt. Scott, along the summit trail.

1. On the image in the placemark balloon, use a protractor to measure the angle of repose of the pumice slope there.

Measured Angle of Repose = ______________

How does your measured angle of repose compare with the angles you calculated at placemarks A and B?

Placemark D. Click the icon for placemark D to fly to it and view the image in the placemark balloon. The view is of Wizard Island, a younger cinder cone within the caldera of Crater Lake.

1. On the image in the placemark balloon, use a protractor to measure the angle of repose of the cinders that make up Wizard Island.

Measured Angle of Repose = ________________

How does your measured angle of repose compare with the angles you calculated at placemarks A and B and measured at placemark C?

Angle of Repose—Location 4: Mt. Bachelor, Oregon.

Placemark A. Open the Mt. Bachelor, Oregon folder and click the icon for placemark A to fly to it.

1. Close the placemark balloon, and use the Ruler tool and elevation data in the Status Bar across the bottom of the display to calculate the angle of repose of the material making up the slope at placemark A (angle of repose = arctan (elevation /distance)).
Work:

Calculated Angle of Repose = _____________

Placemark B. Click the icon for placemark B to fly to it.

1. Close the placemark balloon, and use the Ruler tool and elevation data in the Status Bar across the bottom of the display to calculate the angle of repose of the material making up the slope at placemark B (angle of repose = arctan (elevation /distance)).
Work:

Calculated Angle of Repose = _____________

Angle of Repose—Location 5: Mauna Kea, Hawaii.

Placemark A. Open the The Big Island, Hawaii folder and click the icon for placemark A to fly to it.

1. Close the placemark balloon, and use the Ruler tool and elevation data in the Status Bar across the bottom of the display to calculate the angle of repose of the material making up the slope at placemark A (angle of repose = arctan (elevation /distance)).
Work:

Calculated Angle of Repose = _____________

Placemark B. Click the icon for placemark B to fly to it.

1. Close the placemark balloon, and use the Ruler tool and elevation data in the Status Bar across the bottom of the display to calculate the angle of repose of the material making up the slope at placemark B (angle of repose = arctan (elevation / distance)).
Work:

Calculated Angle of Repose = ______________

Seeing the Bigger Picture

In this exploration you've measured the angle of repose of talus accumulated on several different slopes. Answer the following questions to the best of your ability.

1. What is the range of angles of repose that you measured or calculated in this exploration?

2. What do the slopes with similar angles of repose have in common with one another?

3. List and explain as many factors as you can think of that might explain the variation in the angles you measured and calculated at the various locations.

Exploration 5:
Mass Wasting — Landslides

Slumgullion Slide, Colorado
Gros Ventre Slide, Wyoming

Log on to the *Encounter Earth* site – http://www.mygeoscience.com/kluge – and click the link for the "Exploration 5: Mass Wasting–Slumps and Slides" KMZ file to begin this activity.

Landslides—Location 1: The Slumgullion Slide, Colorado.

Open the "Location 1: Slumgullion Slide" folder and click the icon for the Location 1 placemark to fly to it. Click the icon for this placemark to fly to it, and open and view the placemark balloon. Click the link to the view the air photo of the Slumgullion Slide. You are looking northeast over a 700-year-old landslide that dammed the Gunnison River creating Lake Cristobal.

Turn off the placemark balloon and familiarize yourself with the landscape by clicking on the "Slumgullion view 1" placemark in the Places panel and then clicking the play button to automatically fly through views 1–5.

The gradient, or slope, of a stream bed is found by dividing the change in elevation between two points on the river in meters (the "rise") by the distance between the two points in kilometer (the "run"). (NOTE: for this exercise, you should have your elevation set to display in meters. Do that by clicking Tools, then Options, and setting the "Meters, Kilometers" preference in the 3D view window.)

1. Determine the thickness of the slide debris at placemark C. This can be done by determining the pre-slide gradient of the Gunnison River between placemarks A and B, and from that information determining the elevation of the former stream bed under the present day slide. Start by collecting data to fill in the blanks in Figure 5.1. You can also open placemark C to see a labeled diagram.

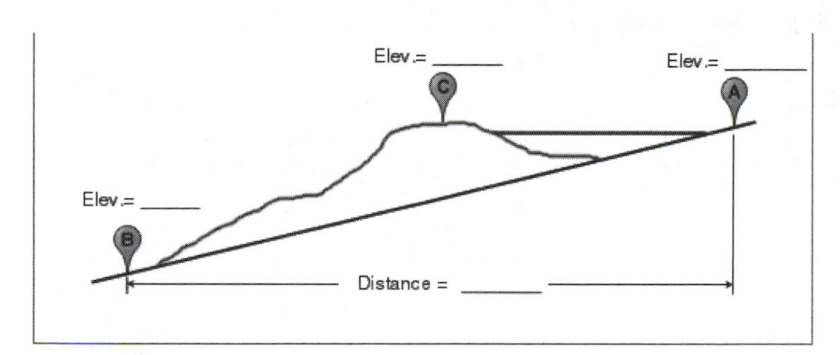

Figure 5.1

Determine the gradient in m/km of the pre-slide Gunnison River between placemarks A and B and report your answer here.

a. Using elevation data displayed in the Status Bar at the bottom of the display, calculate the change in elevation (rise) between placemark A and placemark B.

Rise from A to B = __________m

b. Use the Ruler tool to determine the distance (run) from placemark A to placemark B.

Distance from A to B = _________km

Calculate the pre-slide gradient from A to B.
Work:

Pre-slide gradient = ________________m/km

Devise a way to determine the elevation of the pre-slide riverbed directly under placemark C and report that elevation here (it may help you to label it on Figure 5.1, too).

Calculate the thickness of the slide at location C, and check your calculated thickness with that reported in Chapter 6 of the paper by Varnes, et al. cited in the "Location 1: Slumgullion Slide, Colorado" placemark balloon. (http://pubs.usgs.gov/bul/b2130/)
Work:

Your calculated thickness = _________

Thickness calculated by Varnes, et al. = _____________

2. What term properly describes the landform at placemarks D? _______________.

What is the height of that landform? ______________

Comment on the stability of the area at placemarks D, and cite evidence to support your comments.

Landslides—Location 2: The Gros Ventre Slide, Wyoming

Open the "Location 2: Gros Ventre Slide" folder and click the icon for the Gros Ventre placemark to fly to it. Open and read the placemark balloon.

1. Determine the gradient of the slide area between placemarks A and B.

Change in elevation between A and B = __________m   Distance from A to B = __________km

Gradient of the Gros Ventre Slide area = ____________m/km

2. Click back to the Location 1: Slumgullion Slide placemark, and determine the gradient of the Slumgullion Slide area (between placemark C and either of the placemarks labeled D).

Gradient of the Slumgullion Slide area = ___________m/km

Summary:

1. Provide a qualitative comparison of the volume of material involved in the Slumgullion and Gros Ventre Slides.

2. Compare the rates at which the Slumgullion and Gros Ventre Slides moved.

3. Describe the relationship between the gradients of the slide areas and the speeds at which the slides moved.

Exploration 6:
Groundwater in Karst

Bay County, Florida

Log on to the *Encounter Earth* site – http://www.mygeoscience.com/kluge – and click the link for the "Exploration 6: Groundwater in Karst" KMZ file to begin this activity.

Open the "Groundwater in Karst" folder and click the "?" icon to fly to the study area. Study the geography and topography within the red outlined Study Area.

The landscape here is developed on limestone that, over the years, has been dissolved and carried away by groundwater, leaving a network of caverns under the ground. Where the roofs of those caverns have collapsed, the resulting sinkholes provide a window to the water table below. The level of the water in each sinkhole reflects the height of the local water table.

The Study Area map in Figure 6.1 below also appears on the Places panel in the sidebar. You can turn the map on by checking the box next to it in the Places panel, and you can adjust the opacity up and down to suit your needs.

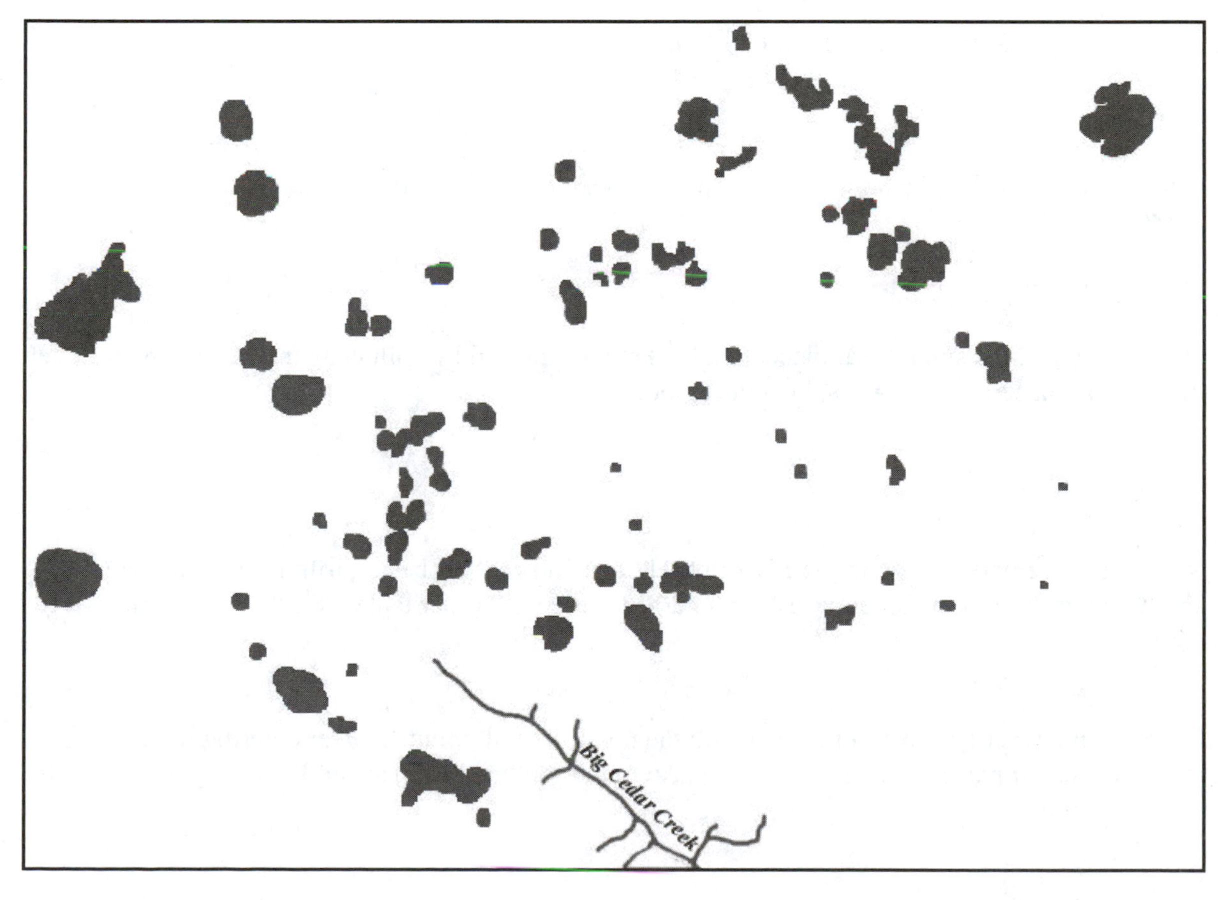

Figure 6.1 Karst Study Area Map

1. Determine the elevation of the surface water in as many of the sinkholes on Figure 6.1 as you can. You will find it helpful to adjust the opacity of the Study Area Map to more easily match the locations on Figure 6.1. Use neat, small print to enter those elevations directly onto Figure 6.1.

a. Mark the general area of the highest water table with an "X" on your map. If there are two areas of high water table, mark them both with an "X."

b. Mark the general area of the lowest water table with a "Y."

c. Using two or three arrows at least 5 cm long, indicate the general direction the groundwater will flow in this area.

d. Attempt to draw the 26, 22, 18, and 14 m contours on the surface of the water table within the study area. Contours should be very smooth and rounded (don't be overly concerned if a few of the smaller sinkholes' water levels do not seem to fit a pattern).

e. Groundwater will flow from the higher elevations to the lower elevations, crossing the water table contours perpendicular to them. Draw three or four short arrows on each contour suggesting the direction the water is flowing.

f. What is the elevation of Big Cedar Creek where it leaves the Study Area?

_________m

What is the source of water for Big Cedar Creek in this area?

2. Double click the placemark A icon in the Places panel. What economic activity is in evidence around the sinkhole there?

a. Would the water in the sinkhole at placemark B likely be affected by pollutants introduced in the water at the sinkhole at placemark A? Explain your reasoning.

3. Double click the placemark C icon in the Places panel. A landowner wishes to drill a water well at placemark C. About how deep will the landowner have to drill to reach the water table? Explain how you arrived at your answer.

a. Should the landowner be more concerned with the possibility of polluted water entering his well from the lake to the west of placemark C, or from the lake to the southeast of placemark C? Explain!

Exploration 7:
Alpine Glaciation — Active Ice

Southeast Coastal Alaska

Log on to the *Encounter Earth* site – http://www.mygeoscience.com/kluge – and click the link for the "Exploration 7: Alpine Glaciation—Active Ice" KMZ file to begin this activity.

Double click the "?" icon for the Location 1: Southeast Coastal Alaska placemark to fly to it. This view presents an overview of the extent of modern glaciation occurring in that area.

1. Double click the icon for placemark A in the Places panel to fly in for a closer look.

a. What evidence can you cite that supports the idea that glacial ice flows as if it is a very thick, viscous plastic?

b. What term describes the dark material at placemark A? ____________________. Fly up to the glacier to find the source of that material, and describe the source of it.

2. Double click the icon for placemark B in the Places panel.

a. Describe the change in width of the tributary glaciers to the right and left of placemark B that occurs as they join the main glacier in the foreground of this view.

b. What are some possible causes for the observed change in the width of these glaciers?

3. Double click the icon for placemark C in the Places panel.

a. What term properly describes the deep fractures in the glacial ice here?__________________. What does the presence of these fractures suggest about the physical properties of the ice here?

b. Explain how a glacier can exhibit properties that suggest both plastic flow and brittle fracture in the same place. Consider the thickness of glacial ice, and the factors that make ice capable of plastic flow.

4. Double click any icon for placemark D in the Places panel. The placemarks D are located at places where there are wide open fractures (or crevasses) in the ice.

a. Study the topography of the valley, and suggest a reason for the opening of those crevasses at each placemark D.

b. Double click the icon for placemark E in the Places panel and compare the width of the crevasses here with those on the main glacier. Explain the differences you observe (the feature at E is called an icefall).

c. To drive the point home, double click the icon for placemark F in the Places panel. Compare and explain the differences in the crevasses above and below the placemark.

5. Double click the icon for placemark G in the Places panel. Describe the event that occurred at this placemark, and describe the material that was added to the glacier by this event.

6. The snowline marks the point on a glacier above which more snow accumulates than melts each year. Below the snowline, there is net melting of the glacier. Often the snowline can be observed as a boundary between new, clean snow and old, dirty snow. Fly around this area, and try to determine the approximate elevation of the snowline in this region.

Approximate elevation of the snowline here = _____________m

7. Double click the icon for placemark H in the Places panel. Why has so much moraine accumulated at this location?

8. Explore the lower-resolution imagery just west of this location, and cite several pieces of evidence that support the notion that glaciation was more extensive in this area in the past.

Exploration 8:
Alpine Glaciation — After the Ice

Galena Mountain and Turquoise Lake, near Leadville, Colorado

Log on to the *Encounter Earth* site – http://www.mygeoscience.com/kluge – and click the link for the "Exploration 8: Alpine Glaciation—After the Ice" KMZ file to begin this activity.

Open the "Location 1: East of Leadville, Colorado" folder and click the icon for the Location 1 placemark to fly to it. The landscape here has recently been modified by alpine glaciation, and in fact a few small glaciers exist at the highest elevations today.

1. Fly to and study each placemark, A through F, in order. Name the glacial landform at each placemark, and briefly describe how it was formed.

A is ____________________ Formation:

B is ____________________ Formation:

C is ____________________ Formation:

D is ____________________ Formation:

E is ____________________ Formation:

F is ____________________ Formation:

2. Double click the icon for placemark G in the Places panel, and turn on the "Turquoise Lake Map" layer and the "Profile from G to H" path by checking the boxes next to them in the Places panel. Refer back to the display as you construct a topographic profile across the Turquoise Lake valley from placemark G to placemark H. Collect your data from the map in Figure 8.1, and draw your profile on the grid below it.

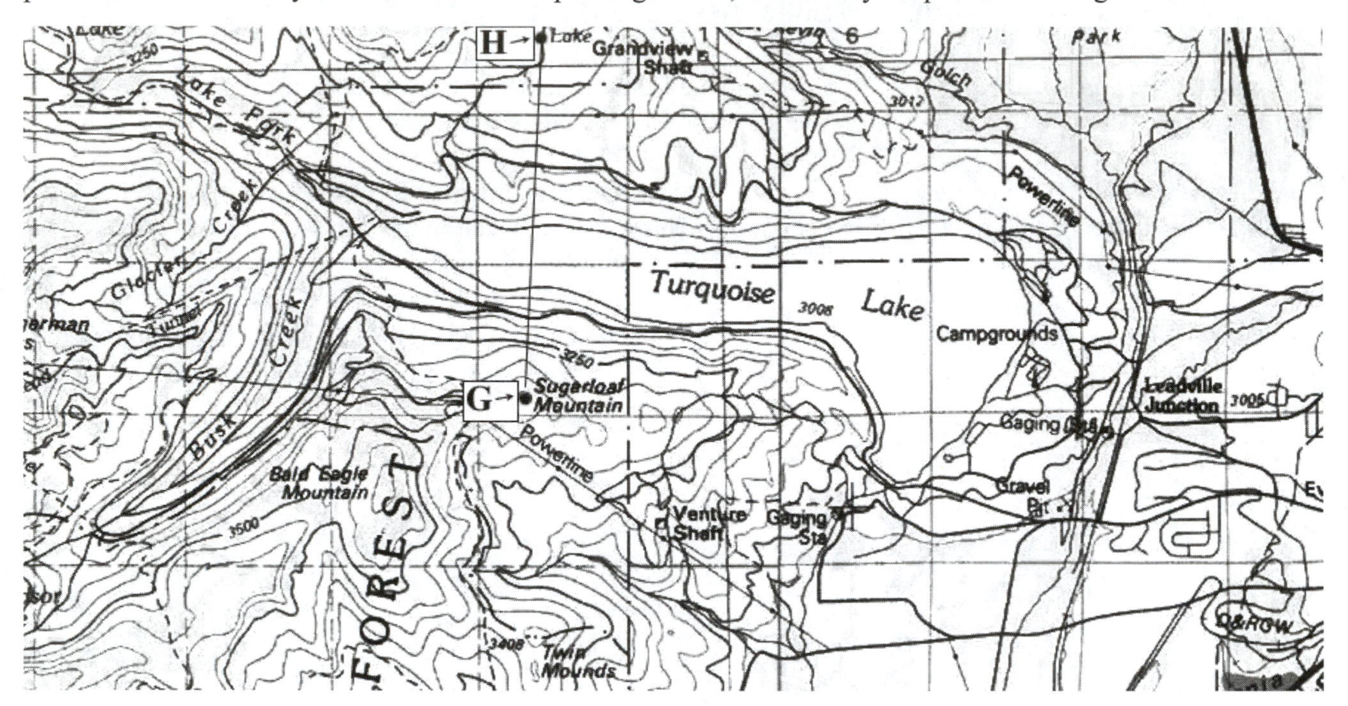

Figure 8.1
(Data compiled by the U.S. Geological Survey)

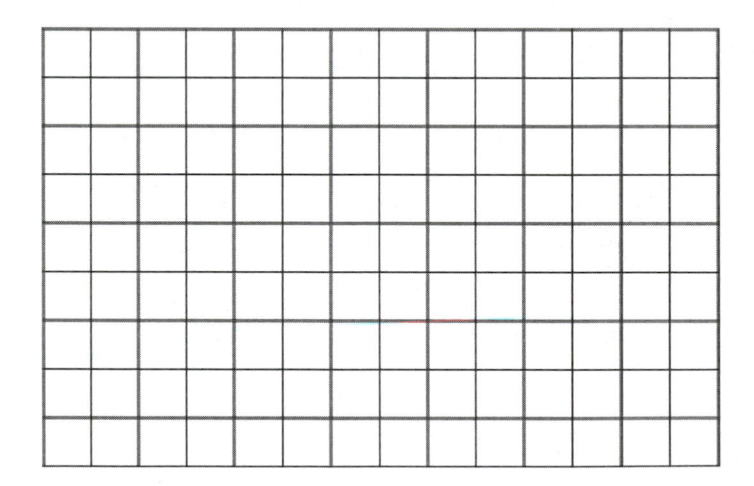

a. What letter of the alphabet does the shape of your profile most nearly resemble? _________

b. Fly back around the area you explored at placemarks A–F and comment on the cross-sectional shape of the glacial valleys in the area.

3. Reduce the opacity of the Turquoise Lake Map layer to about 20%, and turn off the Profile from G to H path. Then double click on any of the placemark I icons in the Places panel. Placemarks I are hovering over cirques carved by ice that flowed off toward the top of your screen (ESE from cirques toward Turquoise Lake).

a. Double click the icon for placemark J to fly to it, and describe the appearance of the landform at placemark J. What is that landform called, and what is the origin and composition of the material it's composed of?

Description:

Name:______________________ Origin and composition:

b. Double click the icon for placemark K to fly to it, and describe the appearance of the land at placemark K. What is that landform called, and what is the origin and composition of the material it's composed of?

Description:

Name:______________________ Origin and composition:

Exploration 9:
Continental Glaciation — After the Ice

Mud Pond, NY
Long Island, NY

Log on to the *Encounter Earth* site – http://www.mygeoscience.com/kluge – and click the link for the "Exploration 9: Continental Glaciation—After the Ice" KMZ file to begin this activity.

Landforms Created by Continental Glaciation—Location 1: Mud Pond, NY

Note: Set the vertical exaggeration to "3" in the Tools > Options > 3D View tab.

Open the "Location 1: Mud Pond, NY" folder and click the "?" icon to fly to the study area. The landscape here is has been modified by Pleistocene continental glaciation, and the elongated hills in this view are composed of unsorted sediments ranging in size from clay to boulders.

1. What is the area covered by the map, in mi^2 ? ______________________________

2. Fly around the map area, zooming in and out as you explore this area. Then make the following measurements and observations.

a. What term properly describes the landforms here? __________________________

b. What is the average length __________ , width ______________ , and local relief _____________ of these landforms?

c. Assuming these landforms were deposited by advancing glacial ice, what compass direction did the ice come from?________________________

d. View the landforms in profile
(double click the icon for placemark A to zoom in to a good starting place).

Are all of the landforms symmetrical in their north-south profiles?

If not, describe the shape of the asymmetrical landforms.

3. Fly to the nearby Bailey Hill location by clicking the overlay icon next to it in the Places panel.

a. How does Bailey Hill compare in length, width, and height with the landforms around Mud Pond?

b. Reduce the opacity of the map to almost zero. How does the presence of this landform appear to affect land use patterns in the area?

Landforms Created by Continental Glaciation—Location 2: Long Island, NY

Open the "Location 2: Long Island, NY" folder and click the "?" to fly to and get an overview of the study area. Then click the icon for the Transect A path to fly to it.

1. Move your cursor along the green transect line, noting the elevation in the Status Bar at the bottom of the display.

a. Click the pushpin icon in the toolbar above the display to create a placemark at the highest elevation along the transect. Write the elevation of the placemark in the Name text box, and click "OK." If there are two areas of high elevation, place a second placemark there.

b. Repeat the steps in a. above at Transect B and Transect C.

c. You have just located the terminal moraine(s) that form the backbone of Long Island, NY. What is the average elevation of the high points of those moraines?____________________________________

2. Double click the icon for placemark A to zoom out to a wide view of the region.

a. Figure 9.1 is a map of Long Island with two terminal moraines indicated. Draw the probable continuation of the two moraines to the eastern end of the map.

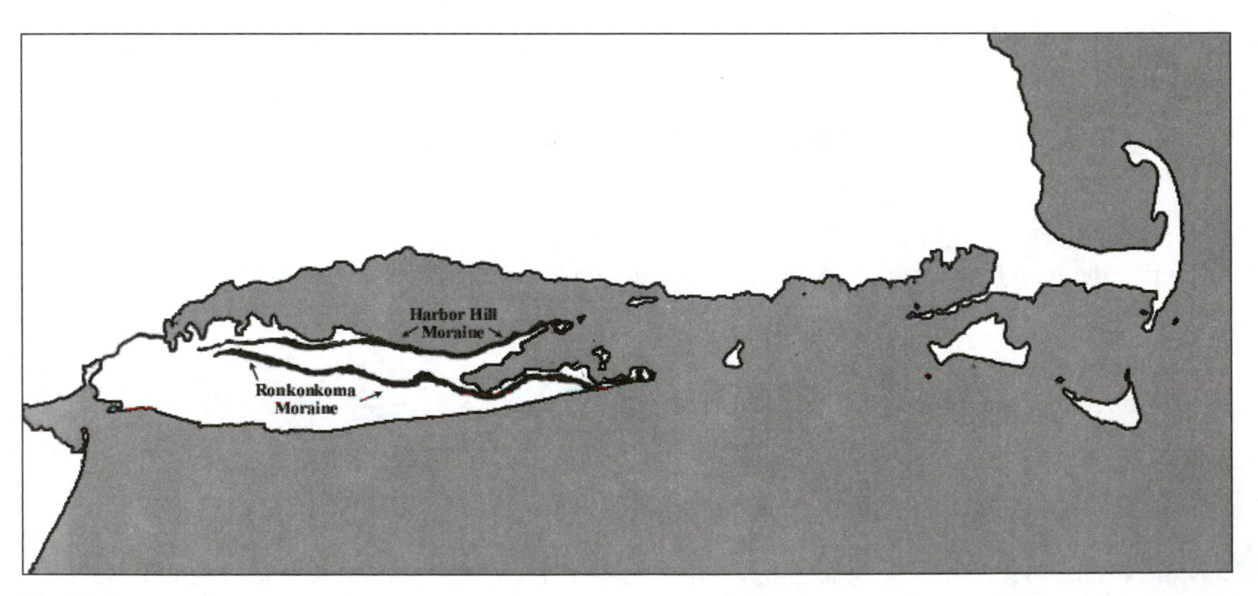

Figure 9.1

Exploration 10: Shorelines

Sandy Point, Kiawah Island, South Carolina
Coastal California, south of Montana De Oro State Park

Log on to the *Encounter Earth* site – http://www.mygeoscience.com/kluge – and click the link for the "Exploration 10: Shorelines" KMZ file to begin this activity.

Location 1: Sandy Point, Kiawah Island, South Carolina

Open the "Shorelines" folder and double click the "?" icon for the Location 1—Sandy Point placemark to fly to it. Carefully study the appearance of the coastal feature that dominates this view. Note the direction of the offshore waves, and the pattern that the breaking waves make on the shoreline. You may want to zoom in and around the area for a closer look, and back out to view the feature in the context of the larger area. You can return to the original view for this question by double clicking the "?" placemark in the Places panel, or directly on the icon on the display. After you have observed and familiarized yourself with this particular landform and the general area, do/answer the following:

1. Is the landform ***erosional***, or ***depositional***? (circle one). What evidence can you provide to support your answer?

2. What term best describes this landform? ________________

Draw your answers to questions 3 and 4 on Figure 10.1, below:

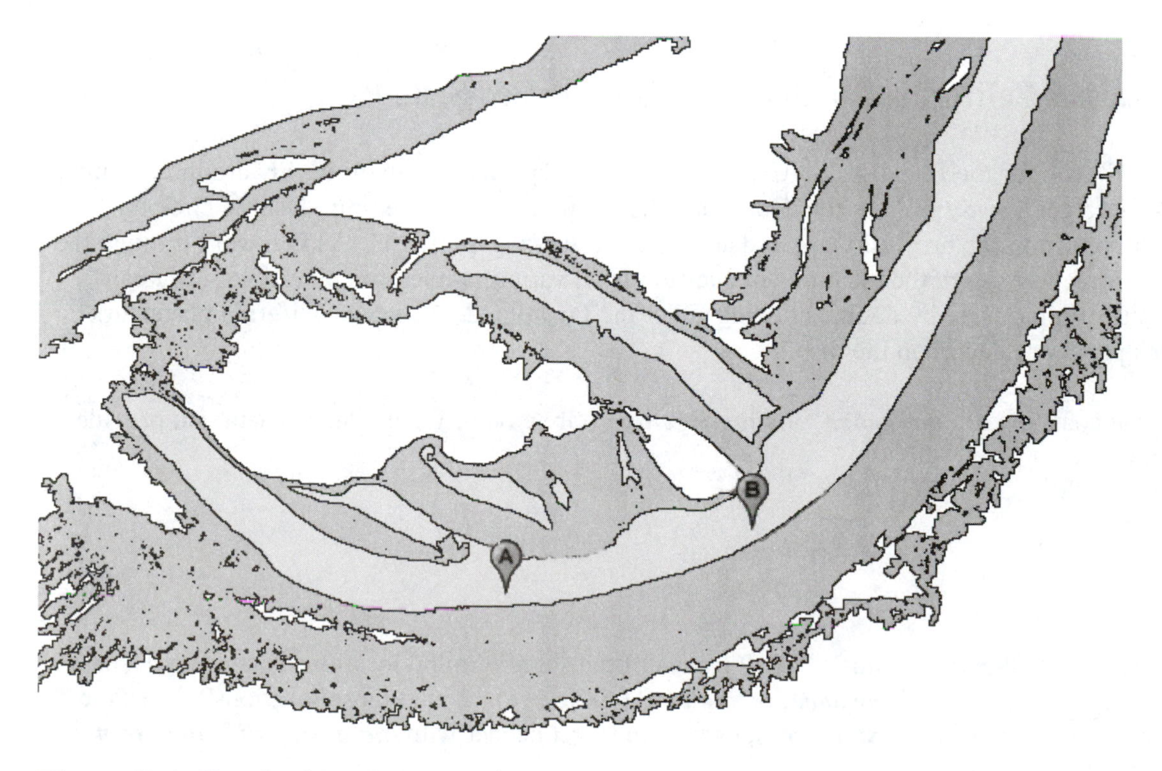

Figure 10.1 Sketch of Sandy Point, Kiawah Island, South Carolina

3. Draw one long arrow to indicate the general direction of the longshore current and beach drift in this area.

4. With an "x", label the most recently deposited sediments visible in the image area.

5. On Figure 10.2, draw several short arrows to indicate the specific path an individual grain of beach sand might travel in the area between placemarks A and B.

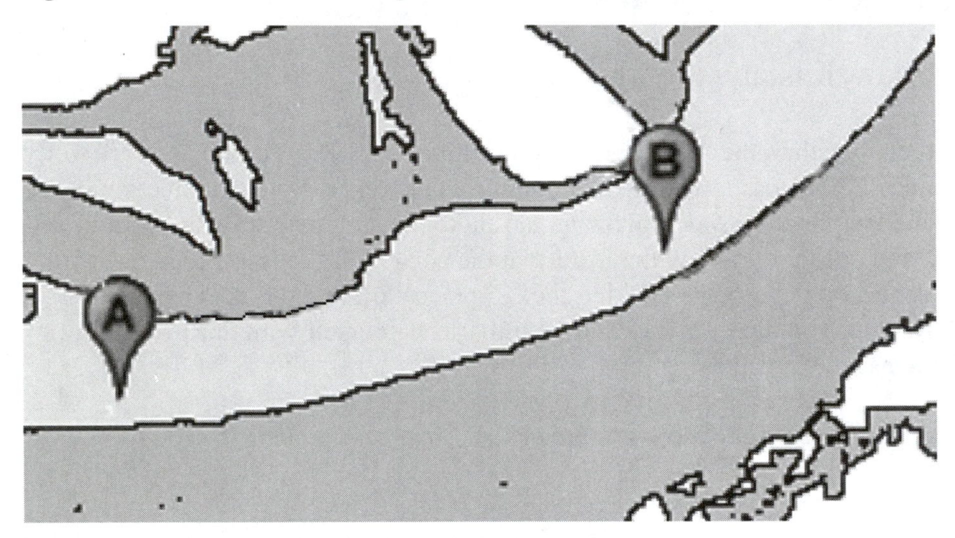

Figure 10.2 Close-up sketch of Sandy Point.

6. Zoom out to an eye altitude of 20 or so miles. What is the most likely source of the sand that this landform is composed of?

Location 2: Coastal California, south of Montana De Oro State Park

Double click the "?" icon for the Location 2—Coastal California placemark to fly to it. Fly around the area, zooming in and out to get a sense of both the details and the big picture of the setting. Double click the "?" placemark again to return to the original view, and answer the following questions. As you work through the questions, take the time to explore the locations in question from various angles and elevations. You can always return to the original view by double clicking either the Location 2—Coastal California placemark icon in the Places panel or its icon on the display.

1. Is the area here a coastline of ***emergence***, or ***submergence***? (circle one). What evidence can you provide to support your answer?

2. Study the appearance of the sea within 100–200 meters of the shore (around locations A and B). Does the depth of the water appear to be ***relatively deep***, or ***shallow***? (circle one) around those locations? Does the water depth appear to be ***relatively constant***, or does it seem to get deeper with ***increasing distance from shore***? (circle one).

3. What term best describes the seafloor at placemarks A and B?

4. What term best describes the landforms at the placemarks labeled D?

5. Describe the topography of the land surface in the areas around placemarks labeled C.

What is the average elevation of those areas? _______________

a. What is the average width of that flat area? (report your answer in a unit that makes sense for an area of that extent)

____

6. a. What is the origin of the spires of rock on either side of placemark E?

b. How are those spires of rock related to the landforms at placemarks D?

7. Double click placemark F. What is the average elevation of the land surface around that placemark?

____

8. What evidence can you site to support the statement that the relatively flat surface at F is older than the surface at C?

9. What term best describes the relatively flat surfaces at A and B, C, and F?

____

10. Describe the tectonic events that combined with the action of waves to produce the landscape in this area. Quantify the extent of the uplift between episodes.

Exploration 11:
Plate Tectonics — Divergent and Transform Boundaries

Red Sea rift zone
San Andreas Fault near Wallace Creek, California

Log on to the *Encounter Earth* site – http://www.mygeoscience.com/kluge – and click the link for the "Exploration 11: Plate Tectonics—Divergent and Transform Boundaries" KMZ file to begin this activity.

Divergent Boundaries

Expand the folder "Divergent Boundaries—Location 1" and click the "Divergent Boundaries—Location 1" placemark. You are looking at the Red Sea between the African and Arabian tectonic plates.

1. Examine the shape of the shoreline of the Red Sea. What evidence can you cite to support the notion that the northeast (right-hand) and southwest (left-hand) shores of the Red Sea were once joined to each other?

2. Click the box next to the path "Red Sea Rift" to turn it on. If you assume that the shores of the lake represent the limits of the forming rift valley, how should the ages of the rocks on the northeastern shore of the Red Sea compare with the ages of the rocks on the southwestern shore?

3. Studies have shown that the rocks rimming the shores of the Red Sea in this area are approximately 20 million years old. If that is the case, is the Red Sea rift opening at the same rate throughout its length?

What evidence can you cite to support your answer?

4. Use the Ruler tool to determine the width of the rift valley at placemark A and placemark B in meters, and convert your measurement to mm. Record your measurements below.

Width of rift valley at placemark A = ____________m = _____________mm

Width of rift valley at placemark B = ____________m = _____________mm

5. Using the estimated age of the rift of 20 million years, calculate the annual spreading rate in mm at placemark A and again at placemark B.

Work:

Placemark	Latitude of Placemark	Spreading Rate at Placemark
A	___________	___________mm/yr
B	___________	___________mm/yr

6. Click open placemark A or B and follow the link to an abstract of an article concerning the spreading rate of the Red Sea rift zone. Describe how closely your measurements come to those reported in the article.

Transform Boundaries

Expand the folder "Transform Boundaries—Location 1."

You are looking at a section of the San Andreas Fault in California. At this location, the beds of several streams and gullies have been offset by displacement along the fault. The red line represents the approximate location of the fault in this area, and placemarks A, B, and C mark the location of offset valleys.

Click placemark A in the sidebar to zoom in to Wallace Creek. Turn on the "Wallace Creek Map" layer by clicking the box next to it. You may want to make it more or less opaque using the slider at the bottom of the Places sidebar.

1. Use the Ruler tool to determine the current offset, in feet, of Wallace Creek where it crosses the San Andreas Fault, and enter that measurement on Figure 11.1, below:

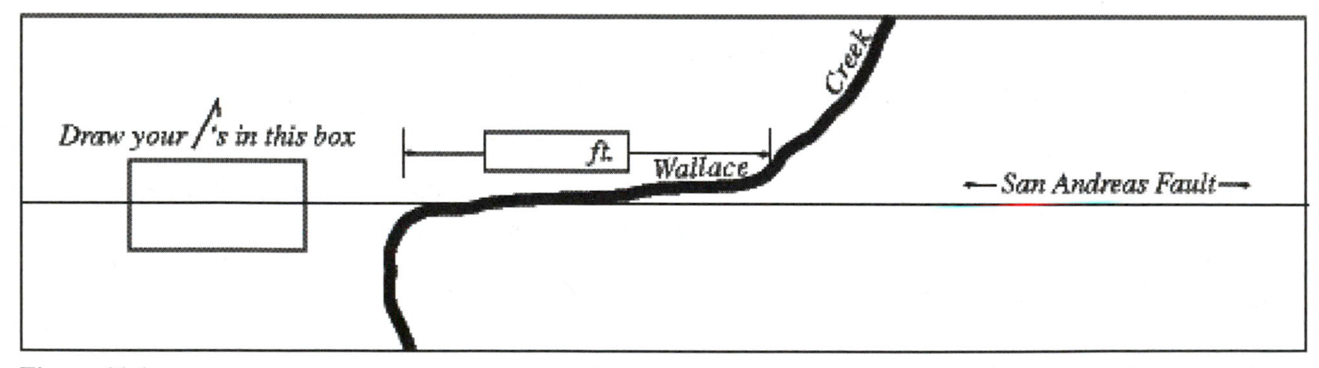

Figure 11.1

2. Imagine you are standing at placemark A and looking NE across the fault. Is the land on the opposite side of the fault moving to your right? Or is it moving to your left?

How would the motion appear to someone standing on the opposite side of the fault looking back toward you?

3. On Figure 11.1, draw a pair of half arrows on either side of the fault indicating the relative directions of movement of the land on either side of the fault. See Figure 11.2.

Figure 11.2

4. Use the Ruler tool to determine the offsets on the smaller river at placemark B, and on the gullies on either side of placemark C. Fill in the chart below:

Wallace Creek offset = ____________ft.

Placemark B offset = ____________ft.

Placemark C offset = ____________ft.

5. Studies indicate that at the time the offset of Wallace Creek began 13,000 years ago, it flowed through the valley indicated at placemark D, and only more recently cut its present channel into the land on the western side of the fault. Determine the average annual offset along this section of the San Andreas Fault (in inches) during the last 13,000 years.

6. Visit the link to the Wallace Creek field trip guide in placemark A (or here: http://www.scec.org/wallacecreek/guides/gsa-wc.pdf), and compare your measured offsets and calculated annual offset rate with those reported in the opening pages of the field trip guide.

Exploration 12:
Plate Tectonics — Convergent Boundaries

Cascadia Margin of the U.S. Pacific Northwest
Pacific Coast of Central America
Aleutian Islands, Alaska

Log on to the *Encounter Earth* site – http://www.mygeoscience.com/kluge – and click the link for the "Exploration 12: Plate Tectonics—Convergent Boundaries" KMZ file to begin this activity.

Convergent Boundaries—Location 1:
The convergent plate boundary along the Cascadia Margin of the U.S. Pacific Northwest

You are looking roughly north along the coast of Oregon and Washington states. Take a few minutes to fly northward from placemark A to placemark B, and then fly back along the spine of the Cascades from Mt. Rainier to Mt. Shasta by clicking on Rainier in the Places sidebar and then clicking the Play button at the bottom of the Places sidebar. After the flight is done, zoom in to the individual volcanoes, and click their icons for pictures and background information.

Study the diagrams and information at placemark A or B, and do the following:

1. Use the Ruler tool to determine the direct distance from the trench to the volcanic peaks immediately to the east. Make several measurements, and record the approximate latitude of your measurement and the trench-volcano distance in the spaces provided below.

Latitude of Observation	Trench-Volcano distance
____________________	______________________
____________________	______________________
____________________	______________________
____________________	______________________

Research has shown that melting begins to occur as the subducted plate arrives at a depth somewhere between 100 and 120 km below Earth's surface. If the less dense molten rock rises to the surface, it may erupt as a volcano.

2. Refer to the diagram and formula described in placemark A or B, and the average trench-to-volcano distance you recorded above, to determine the descent angle of the Juan de Fuca plate beneath the North American continent.

Work:

Descent angle of the Juan de Fuca plate = __________

Convergent Boundaries—Location 2:
The convergent plate boundary along the coast of Central America

Click each of the volcanoes in the Places sidebar to fly from one to the next. Click their icons for pictures and background information. Answer/do the following:

1. Assume that melting along the subducted plate begins between 100 and 120 km below the surface of Earth, and use the diagram and formula in placemark C or D to determine the descent angle of the Cocos Plate beneath Central America.

Work:

Descent angle of the Cocos Plate = __________

Convergent Boundaries—Location 3:
The convergent plate boundary along the Aleutian Islands

Click each of the volcanoes in the Places sidebar to fly from one to the next. Click their icons for pictures and background information. Answer/do the following:

1. Once again, assume that melting along the subducted plate begins between 100 and 120 km below the surface of Earth, and use the diagram and formula in placemark E or F to determine the descent angle of the Pacific Plate beneath the Aleutian Islands.
Work:

Descent angle of the Pacific Plate = __________

Putting It All Together

1. What factors might influence the descent angle of a subducted plate?

2. How does the temperature of a given slab of seafloor affect its density?

3. How does the thickness of a given slab of seafloor affect its weight?

4. What happens to the density and thickness of a given slab of seafloor as it moves away from the rift zone where it originated? Why?

5. Use the Ruler tool to determine the distance from the rift zone where the seafloor originated to the trench where it is being subducted at each of the three locations studied in this exploration.*

Location	Distance from rift/origin
Cascadia Margin	____________________
Coast of Central America	____________________
Aleutian Islands	____________________

*If you are having trouble identifying the origin region, you can turn on the Hint at each location by checking the box next to it in the Places sidebar.

6. Describe and explain the relationship you've observed between the age of a subducted plate, and the descent angle of the subduction of that plate.

Exploration 13:
Volcanism — Global Distribution of Volcanoes

Log on to the *Encounter Earth* site – http://www.mygeoscience.com/kluge – and click the link for the "Exploration 13: Volcanism—Global Distribution of Volcanoes" KMZ file to begin this activity.

In the Layers panel, expand the "Gallery" layer and turn on the "Volcanoes" layer. Once the Volcanoes layer is turned on, you can hide the Layers panel.

Open the "Volcanism—Global Distribution of Volcanoes" folder and double click the "?" icon to fly to Central America. Open this "Global Distribution of Volcanoes" placemark and follow the instructions there to download and open this file in the Google Earth™ mapping service: http://bbs.keyhole.com/ubb/download.php?Number=286222. Open the "Plate boundary model" folder and turn off the "Rotation poles" layer by unchecking the box next to it.

Note: Convergent plate boundaries are drawn in blue, while divergent boundaries are drawn in red on the sea floor and in yellow on land.

Quickly navigate around the world, paying attention to the tectonic plate names and boundaries. Notice that the volcanoes that are displayed vary with the level of zoom.

1. Double click the icon for placemark A in the Places panel.

a. Describe the location of volcanoes with respect to the plate boundaries in the area around placemark A.

b. What is "different" about the volcanoes located around placemark I?

2. Double click the icon for placemark B in the Places panel.

a. Once again, describe the distribution of volcanoes with respect to the location of plate boundaries.

3. Double click the icon for placemark C in the Places panel.

a. Describe the distribution of volcanoes with respect to the location of plate boundaries here as well. Then double click the icon for placemark D, and describe the distribution of volcanoes in that view.

b. Zoom out so that placemarks A, B, C, and D are all visible (or double click the icon for the "Pacific View" placemark). Then zoom in and fly along the rim of the Pacific Ocean. Explain why the Pacific Rim is called the "Ring of Fire," and describe the conditions that exist to create the Ring of Fire.

4. Double click the icon for placemark E and describe/name the type of plate boundary in this area.

a. Describe the distribution of volcanoes with respect to this type of plate boundary. How does the distribution of the volcanoes differ from the location of the volcanoes you observed at placemarks A, B, C, and D?

5. Double click the icons for placemarks F, G, H, and I in succession, stopping to study the distribution of the volcanoes in each of these areas.

a. How does the distribution of volcanoes in each of these areas differ from those viewed previously? What do the distributions of the volcanoes in each of these places have in common with one another?

b. The volcanoes at placemarks F, G, H, and I are all "hot spot" volcanoes. Search the globe, and report the latitude and longitude of at least two additional probable hot spots.

Exploration 14:
Volcanism — Lava Flows

Davis Lake, Oregon

Log on to the *Encounter Earth* site – http://www.mygeoscience.com/kluge – and click the link for the "Exploration 14: Volcanism—Lava Flows" KMZ file to begin this activity.

Open the "Location 1: Davis Lake, Oregon" folder and click the icon for the Location 1 placemark to fly to it. Open and view the placemark balloon.

Make sure the "Davis Lake Flow Map" overlay and "Sample Area" polygon are turned off in the Places panel. Then fly around, zoom in and out, to carefully explore the entire lava flow.

1. Double click the icon for placemark A in the Places panel to fly to it. Placemarks A and B indicate the location of cinder cones.

a. Which one is the older of the two flows? Provide a rationale for your answer.

2. Turn on the Davis Lake Flow Map overlay and use the opacity slider to set the opacity at about 30%. Then double click the icon for placemark C in the Places panel.

a. Study the flow at placemark C, and trace the outline of the flow on the map, Figure 14.1, below. You may want to double click the Davis Lake Flow Map layer and adjust the opacity up and down as you make your drawing.

b. Indicate with a "C" on the map the apparent source of the lava of the flow at placemark C.

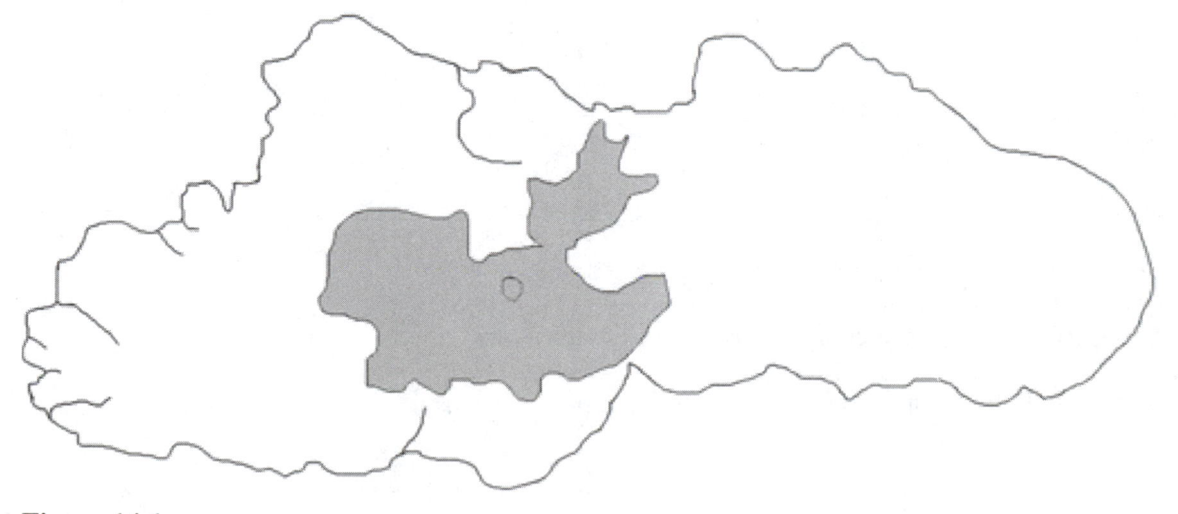

Figure 14.1

c. Decrease the opacity of the map, and zoom out to view the entire flow. Then, on Figure 14.1, trace the outline of as many individual flows as you can find.

d. On Figure 14.1, indicate with arrows the direction of flow within each flow you've traced.

e. Finally, describe how you might be able to determine the relative ages of each flow, and indicate the age of each flow you've traced by numbering them in order from oldest to youngest on Figure 14.1.

3. Determine the thickness of the flow at placemark D by comparing the elevation of the flow at placemark D with the elevation of the land surface at placemark E.

Thickness of flow at placemark D = ________________

4. Determine the thickness of the flow at placemark F by comparing the elevation of the flow at placemark F with the elevation of the land surface at placemark G.

Thickness of flow at placemark F = ________________

5. What might account for the variation in the thickness of the flow observed between placemarks D and F?

6. In the Places panel, turn off the Davis Lake Flow Map overlay, and turn on the Sample Area polygon.

a. Using methods similar to those used in questions 3 and 4 above, determine the average thickness of the flow in this area. Report your answer in meters. (You may have to reset the units in the Status Bar to meters. Click Tools, then Options to reset the elevation units in the 3D View tab.)

Average thickness of the flow within the Sample Area polygon = ______________m

b. Use the Ruler tool to make the measurements necessary to calculate the surface area of the Sample Area polygon.

Measurements made: ____

Surface area of Sample Area polygon = _____________m^2

c. Calculate the approximate volume of the lava in the Sample Area.

Volume of lava in the Sample Area = _____________ m^3

Exploration 15:
Volcanism — Calderas

Crater Lake, Oregon
Newberry Volcano, Oregon

Log on to the *Encounter Earth* site – http://www.mygeoscience.com/kluge – and click the link for the "Exploration 15: Volcanism—Calderas" KMZ file to begin this activity.

Calderas—Location 1: Crater Lake, Oregon.

Open the "Location 1: Crater Lake" folder and click the volcano icon for the Crater Lake placemark to fly to it. Open and view the placemark balloon. Turn on the "Former Summit of Mt. Mazama" placemark in the Places panel, and then double click the "?" icon to fly to it, and open and read the placemark balloon there. Double clicking the red cross icon for the "Former Summit" placemark will fly you to another good perspective to visualize the former Mt. Mazama.

Note: You may want to turn on and off the various placemarks in the Places panel to provide unobstructed views as you work through this exercise.

1. Double click the icon for placemark A in the Places panel to fly to it. Use the Ruler tool to collect the data needed to calculate the surface area of Crater Lake in mi^2.

a. Diameter of Crater Lake = _______________

Surface area of Crater Lake = ____________________

2. Double click the icon for placemark B and then for placemark C. You may want to fly back and forth between them as you answer the following questions.

a. Describe the cross-sectional shape of the valleys indicated by the placemarks.

b. These U-shaped valleys are characteristic of valleys formed by glaciers. How does the presence of these valleys help to indicate the former presence of a large volcanic mountain where Crater Lake now exists?

3. Turn on the Crater Lake Bathymetry Map and Crater Lake Map Key. As with most geologic maps, the oldest units in the key are located at the bottom right, and the most recent at the top left.

a. Where on the map did the most recent volcanic activity occur?

b. What type of geologic activity dominates the crater today?

Calderas—Location 2: Newberry Volcano, Oregon.

Open the "Location 2: Newberry Volcano" folder and double click the "?" icon to fly to it. Then double click the volcano icon for the Newberry placemark to fly to it. Open and view the placemark balloon. The floor of the caldera contains several volcanic features that underscore the volcanic activity in the area.

1. Describe and name the volcanic landforms at each of the following placemarks.

a. A

b. B

c. C

d. D

2. Figure 15.1 is a map of the field of view at placemark A. Draw several arrows to indicate the direction(s) of lava flow within the map area, and mark the location of the origin of the flow with the letter "X."

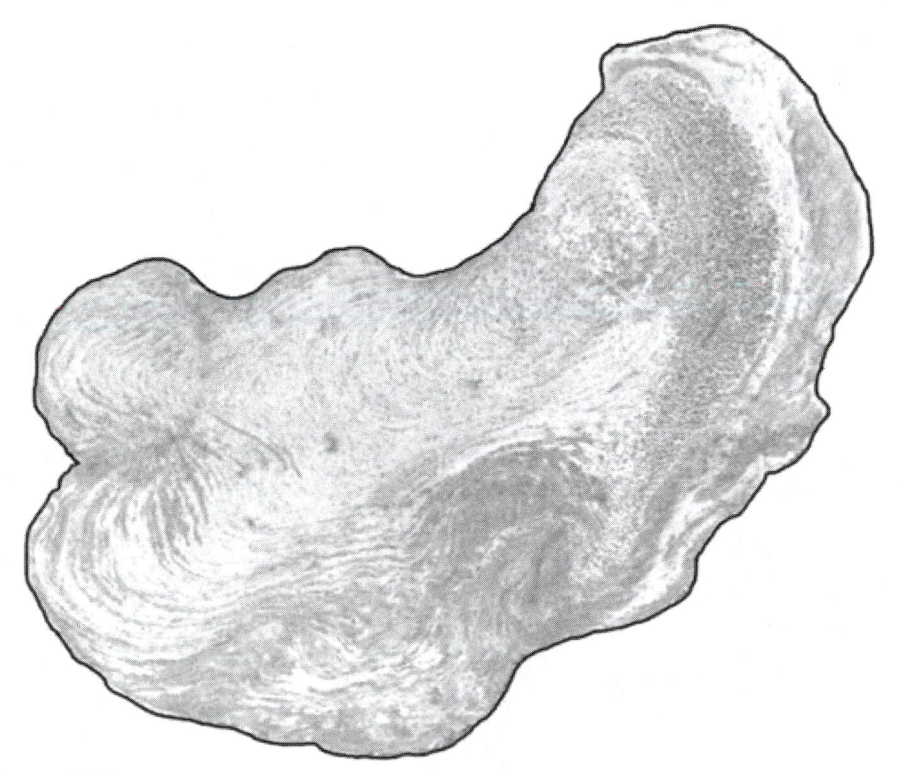

Figure 15.1

3. Compare the size of the Newberry caldera with that of Crater Lake.

4. List the features the floor of the Newberry caldera has in common with the floor of the Crater Lake caldera.

5. Other than the obvious lack of lake filling the caldera, in what ways is the Newberry caldera different from the Crater Lake caldera?

Exploration 16:
Structural Geology — The Boulder Flatirons

West of Boulder, Colorado

Log on to the *Encounter Earth* site – http://www.mygeoscience.com/kluge – and click the link for the "Exploration 16: Structural Geology – The Boulder Flatirons" KMZ file to begin this activity.

Double click the "?" icon for the "Location 1: The Boulder Flatirons" placemark to fly to it. The view is to the south along the Colorado Front Range of the Rocky Mountains. The Flatirons were formed when the Permian/Pennsylvanian-aged sedimentary rocks of the Fountain Formation were tilted upward as the older rocks to the west rose during the Laramide Orogeny about 70 million years ago. Subsequent erosion has reduced those tilted slabs of the Fountain Formation to large, triangular slabs of rock that seem to lean up against the older rocks to the west of them.

Open the Location 1 placemark in the display, and click the link to the Map Key to open it in a separate browser window. Keep the key open to for reference as you work through this exercise. Note that the Map Key can also be displayed as an overlay.

Double click the "Boulder Area" overlay to turn on the geologic map of the area. Fly in for a closer look, and adjust the opacity of the map up and down to get a feel for how the geology of the area affects the appearance of the land. When you are finished browsing, set the map opacity to about 40%.

1. Double click the icon for either placemark A or B in the Places panel. Use the Ruler tool and the elevation data in the Status Bar at the bottom of the display to make the following measurements:

a. Elevation of placemark A = ____________ Elevation of placemark B = ________________

Change in elevation between placemarks A and B = _______________

b. The distance from placemark A to placemark B = _____________

(**NOTE:** It does not matter what unit the measurements above are made in, as long as they are all made in the same unit.)

c. The dip angle of the Fountain Formation can be found by the following formula:

Dip Angle = arctan (change in elevation/distance)

Calculate and record the dip angle of the Fountain Fm in this area. ___________________

2. Double click the icon for either placemark C or D in the Places panel. Notice the fault between those two placemarks.

a. In what direction does the fault here strike? ______________________

3. Assuming the Fountain Fm was originally deposited as a single, thick layer of sandstone and conglomerate rock, and looking at the displacement of the Fountain Fm on either side of the fault, answer the following:

a. To an observer looking across the fault from placemark D to placemark C, did the rocks at placemark C move to the ***right*** or to the ***left*** (circle one) relative to the rocks at placemark D?

b. Did the faulting here serve to ***lengthen*** or ***shorten*** (circle one) the Front Range mountains in this area? Explain your reasoning.

4. Double click the icon for placemark E in the Places panel. Use the map key to determine the age and name of the rock unit that outcrops at placemark E.

Age = ________________ Name = ____________________

a. Are the rocks at placemark E older or younger than the Fountain Fm?

b. How does the dip of these rocks compare with the dip of the Fountain Fm?

Does that suggest that the formation of the rocks at E ***predates*** or ***postdates*** (circle one) the deformation that raised the Fountain Fm? Explain.

5. Turn on the "South of Boulder" overlay and set the opacity to 100%. Double click the icon for placemark F in the Places panel.

a. Using the map key, answer the following:

What is the name of the formation at placemark F? ____________________

And what is it composed of?

b. What visual evidence suggests that the Qrf is younger than the Kp and Kl formations?

c. What is the source area of the Qrf alluvium?

6. Turn on the "Fox Hills Sandstone" overlay and set the opacity to about 30%. Double click the icon for placemark G in the Places panel. The Fox Hills Sandstone is a thin layer of sandstone that dips vertically in this area.

a. Turn off all the map layers and fly north along the outcrop of the Fox Hills Fm. Locate the Fox Hills outcrop at latitude 39° 55' North. Then click the South of Boulder map back on, and note the outcrop on the map. Finally fly back south along the outcrop using the map as a guide.

b. Explain why the outcrop of the Fox Hills Fm on the ground and on the map is discontinuous.

Exploration 17:
Geologic History — Dinosaur Ridge

Dinosaur Ridge, Colorado

Log on to the *Encounter Earth* site – http://www.mygeoscience.com/kluge – and click the link for the "Exploration 17: Geologic History—Dinosaur Ridge" KMZ file to begin this activity.

The Jurassic and Cretaceous—Location 1: The Dakota Hogback

Open the folder "Geologic History—Dinosaur Ridge, Colorado."

1. Double click the icon for placemark A to fly to it, and open the placemark balloon. The view is of the Interstate 70 road cut through the Dakota Hogback, a Jurassic-Cretaceous age formation of sedimentary shales and sandstones that dips steeply to the east. The gray and purple rocks on the western side of the outcrop are largely shales and mudstones. The tan and gray rocks on the eastern side of the outcrop are largely sandstones. The rocks here have not been overturned.

a. Use an "X" to label the oldest rocks on Figure 17.1, and label the youngest rocks with the letter "Y."

b. What "law" of relative dating did you apply to determine the relative ages of the rocks here? Explain that law.

Law ___________________________

Explanation:

Figure 17.1 The Dakota Hogback

c. What evidence is there to suggest that these rocks have been uplifted and deformed?

What law of relative dating did you apply in citing that evidence?______________________

Explain that law:

d. Where do the rocks that are least resistant to weathering and erosion outcrop on Figure 17.1?

What rock types make up that part of the outcrop?

e. The Jurassic rocks here are gray and purple shales and mudstones, while the Cretaceous rocks are tan and light gray sandstones. On Figure 17.1, draw a line representing the approximate Jurassic-Cretaceous boundary, and label the Jurassic- and Cretaceous-aged rocks on either side of the boundary.

f. What is the age, in millions of years, of the boundary you drew here?

____

2. Double click the icon for placemark B to fly to it, and open the placemark balloon. The image is of a cross section of the Cretaceous-aged rocks exposed on the west-facing flanks of the Dakota Hogback.

a. Describe the sequence of events that must have occurred in order for these footprints to have been made, preserved, and exposed to our view. Include in your description what you imagine the environment was like as the "bronto" made these imprints.

b. As noted earlier, the Jurassic rocks here are gray and purple shales and mudstones, while the Cretaceous rocks are tan and light gray sandstones. Use the elevation data in the Status Bar at the bottom of the display to determine the elevation of the Jurassic-Cretaceous boundary in the area around placemark B.

Elevation of the Jurassic-Cretaceous boundary: ______________________

3. Double click the icon for placemark C to fly to it, and open the placemark balloon. The right-hand image is of a bedding plane on a layer of the Cretaceous-aged "Dakota Sandstone" here.

a. Describe the sequence of events that must have occurred in order for these ripple marks to have been made, preserved, and exposed to our view.

4. Double click the icon for placemark D to fly to it, and open the placemark balloon. The image is of a "dinosaur trackway"—a set of many footprints preserved on a bedding plane of a layer of the Cretaceous-aged "Dakota Sandstone" here.

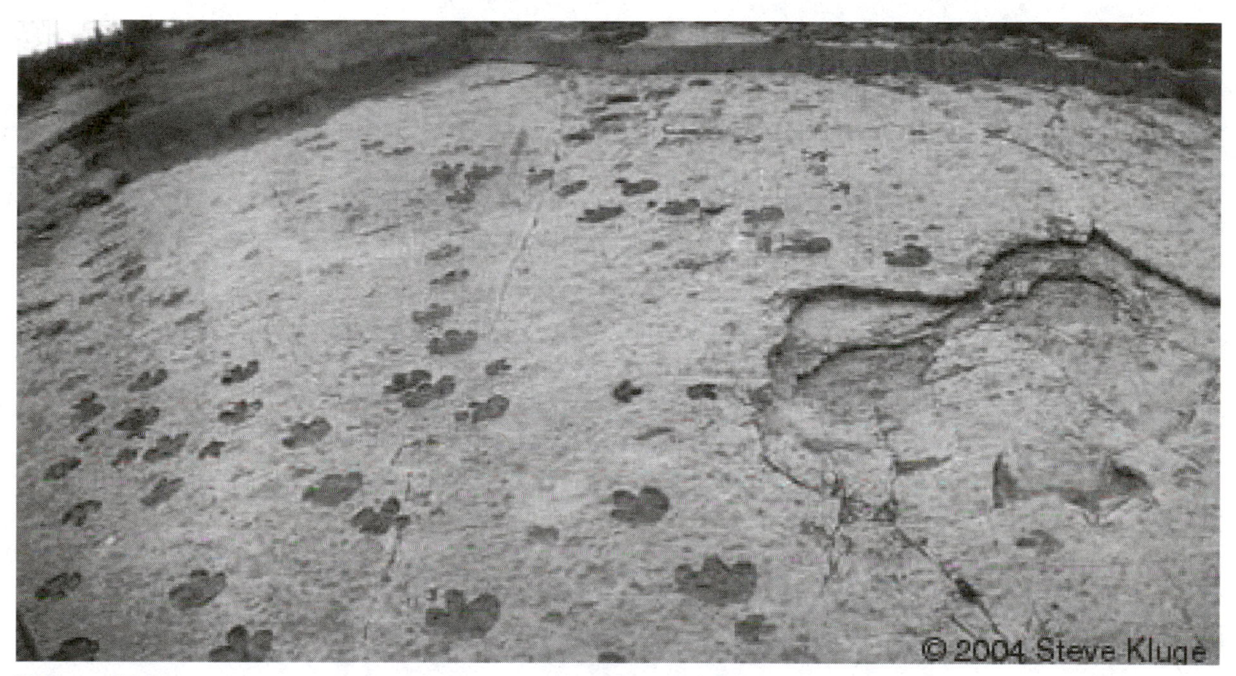

Figure 17.2

a. Two species of dinosaurs left their tracks here: one with a broad, three-toed foot (there were adult and young individuals here, leaving large and small footprints, respectively), and the other with a smaller, three-toed, birdlike foot. Circle a pair of the smaller, birdlike prints on Figure 17.2.

b. On Figure 17.2, draw several arrows indicating the directions that individual animals were moving in this area.

c. What span of time do you suppose is represented by this trackway? Explain your answer.

Exploration 18:
Geologic History — Red Rocks

Red Rocks, Colorado

Log on to the *Encounter Earth* site – http://www.mygeoscience.com/kluge – and click the link for the "Exploration 18: Geologic History—Red Rocks" KMZ file to begin this activity.

Precambrian through Quaternary—Location 1: Red Rocks

Open the folder "Geologic History—Red Rocks."

1. Double click the icon for "Image 1" to fly to it, and open the placemark balloon. Click through to the Earth Science Picture of the Day to read about the geology of the area. Open the balloon for "Image 2" for another labeled image of the same area. You will want to refer back to these images as you work through this Exploration.

a. Use the labels on Image 1 to identify by name the rock that is:

Most resistant to weathering and erosion: ____________________

Least resistant to weathering and erosion: ____________________

Explain the reasoning you used to make those inferences.

b. What is the possible age of the Ralston Creek Fm? ____________________

Explain the reasoning you used to make that inference.

2. Turn on the "Dinosaur Ridge and Red Rocks" map overlay and set the opacity to about 30%. Turn on the "Map Key" overlay as well, and set the opacity to 100%. You can refer to the Map Key by double clicking its icon in the Places panel.

a. Double click the icon for placemark A to fly to that view. Generally, in what colors are the Quaternary rocks drawn?

____

b. What colors are used to represent the Permian- through Jurassic-aged rocks here?

____

c. How does the pattern of Quaternary deposits indicate that they are younger than the Permian-Jurassic rocks in this area?

d. Imagine that you could remove the Quaternary deposits to reveal the rocks beneath them. Describe the outcrop pattern of those rocks in this area.

e. Where do the oldest rocks in this area outcrop?

How old are they?

3. Double click the icon for "Quaternary Landslide Deposits" to fly to it. Note the location of the landslide deposits, and then reduce the opacity of the Dinosaur Ridge and Red Rocks map to near zero. You may want to adjust the opacity up and down as you answer the following question.

a. On the small map of the Qls in Figure 18.1, indicate the "toe" of the landslide with the letter "X" and the source region of the slide material with the letter "Y." You may want to fly around the area and zoom in and out to view the area at various scales and angles.

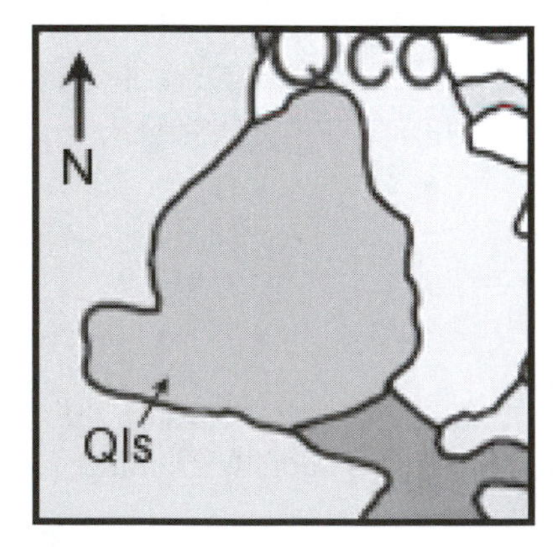

Figure 18.1

Exploration 19:
Natural Hazards

Mt. St. Helens, Washington
La Conchita, California
La Jolla, California

Log on to the *Encounter Earth* site – http://www.mygeoscience.com/kluge – and click the link for the "Exploration 19: Natural Hazards" KMZ file to begin this activity.

Natural Hazards—Location 1: Mt. St. Helens, Washington

Open the "Location 1: Mt. St. Helens Lahar" folder and click the icon for the "Mt. St. Helens Lahars" placemark to fly to it. Open the placemark balloon. Click through to the lahar images linked there.

Click and fly to the icon for placemark A, and open placemark A's balloon. The image is of the Toutle River valley lahar, taken 24 years after the eruption.

Click and fly to the icon for placemark B, and view the close-up image of the lahar in that placemark.

Click and fly to the icon for placemark C and click through to the web page detailing efforts to control sediment flow in the downstream reaches of the Toutle River.

1. Describe the man-made structure at placemark C, and explain how it works and why it was built.

Use the Search panel to fly to Orting, Washington.

2. Notice that Orting is located at the confluence of two rivers. The river NE of town is the Carbon River, the river to the SW is the Puyallup River. Trace both of those rivers upstream, and describe the location and source of water for those streams.

3. What hazards would the inhabitants of Orting face in the event of an eruption of Mt. Rainier (25 miles SE of Orting), and what plans can be made or actions can be taken to prevent loss of life and property if an eruption occurs?

4. Search for Carbonado, Washington, and explain why the inhabitants of Carbonado, though closer to Mt. Rainier, are in less immediate danger of inundation by a lahar if Mt. Rainier were to erupt.

5. Compare the appearance of the Puyallup and Carbon River valleys around Orting with the Toutle River valley at placemark A. How are the valleys similar? And how are the valleys different?

6. Describe any evidence you see that suggests the Puyallup and Carbon River valleys have been the sites of lahars in the past.

Natural Hazards—Location 2: La Conchita, California

Open the "Location 2: Coastal California" folder and click the icon for the "Location 2: La Conchita Landslide" placemark to fly to it, and open the placemark balloon. Click through to view the video (works best in Internet Explorer) and read the description of the January, 2005 event.

1. What geologic conditions existed in the area prior to the 2005 slide that pre-disposed the area to future slides?

2. What were the immediate causes of the 2005 slide?

3. Determine the gradient between placemarks A and B (on the slide itself) and between placemarks C and D on the intact slope adjacent to the slide.

Gradient on the slide = ____________ Gradient on the intact slope = _____________

4. Describe how the occurrence of the slide helps to stabilize the material that has failed during the slide.

Natural Hazards—Location 3: La Jolla, California

Open the "Location 3: La Jolla Landslide" folder and click the icon for the "Location 3: La Jolla Landslide" placemark to fly to it, and open the placemark balloon. Click through to the slide show, and read the article regarding the slide.

Click on the "Soledad Mt. Road Map" overlay in the Places panel, and set the opacity slider to about 25%. Then fly around the area to familiarize yourself with the level of development and the topography of the land. Turn on the "Roads" layer in the Layers side panel.

1. Find two or three other areas (within 0.75 miles of the Location 3 placemark), and list the names of the streets in the area that you think might be subject to damage by landslide in the future. For each location you cite, describe why you think the area is vulnerable.

Summary:

1. Compare the La Conchita event with the La Jolla slide with regard to a) pre-slide warning signs, b) the future threat to property and human life, c) the speed at which the events occurred, and d) the geologic setting of the slides.

2. What actions might local governments in each community take to ameliorate the effects of future events, or to eliminate a recurrence entirely?

Exploration 20:
Impact Structures

Manicouagan, Canada
Barringer Meteor Crater, Arizona

Log on to the *Encounter Earth* site – http://www.mygeoscience.com/kluge – and click the link for the "Exploration 20: Impact Structures" KMZ file to begin this activity.

Location 1: Manicouagan, Canada

Open the "Location 1: Manicouagan Impact" folder and click the "?" icon for the Manicouagan Impact Structure placemark to fly to it. Open and view the placemark balloon, and click and follow the link to some brief background information on Manicouagan.

1. Right click the "Manicouagan" shape icon, and select "Properties" (on a PC) or "Get Info" (on a Mac).

2. Select the "Altitude" tab, and use the slider to raise and lower the shape relative to the land surface. Notice how it blocks the view of the land "under" it, while it leaves the higher land in view above the shape.

Finally, set the Altitude to 450m, and notice the circular ridge beyond, or outside, of the lake. That ridge is actually the rim of the crater! Determine the diameter of the crater in miles, and use that information to determine the area of the crater in mi2.

Area of Manicouagan Impact Structure = ____________________ mi2

Location 2: Barringer Meteor Crater, Arizona

Open the "Location 2: Barringer Crater, Arizona" folder and click the "?" icon for the Barringer Meteor Crater placemark to fly to it. Open and view the placemark balloon, and click and follow the link to some brief background information on the Barringer crater.

1. Visit GPS Visualizer (http://www.gpsvisualizer.com/kml_overlay) to get a higher-resolution image of this area.

a. Once at the site, enter the following in the form:

In the "Enter a center point:" area

Latitude = 35.027
Longitude = –111.022
Enter 4 km for both Height and Width

In the "General options" area

Coordinate input method: Enter a center point
Ground Overlay type: Dynamic
Size of overlay: 2000 pixels
GE document name: Barringer Image
Dimensions: Square
Map: US: USGS aerial photo (1m; 2000px)

2. Measure the diameter of Barringer Crater and calculate the surface area, in mi^2, of the crater.

a. Surface area of the crater = ____________________ mi^2

b. How many times larger is the Manicouagan structure than this one?

3. Study the inside walls of the crater.

a. What evidence exists to suggest that the depth of the crater is less today than it was shortly after it formed 40,000 years ago? (Hint: What process is apparent on the walls of the crater?)

b. Explain why and how this crater will eventually be "erased" from the surface of the Earth.